Date Due

Cram101 Textbook Outlines to accompany:

Technical Communication

John M. Lannon, 11th Edition

Learning System

Cram101 Textbook Outlines is a learning system. The notes in this book are the highlights of your textbook, you will never have to highlight a book again.

How to use this book. Take this book to class, it is your notebook for the lecture. The notes and highlights on the left hand side of the pages follow the outline and order of the textbook. All you have to do is follow along while your instructor presents the lecture. Circle the items emphasized in class and add other important information on the right side. With Cram101 Textbook Outlines you'll spend less time writing and more time listening. Learning becomes more efficient.

Cram101.com Online

Increase your studying efficiency by using Cram101.com's practice tests and online reference material. It is the perfect complement to Cram101 Textbook Outlines. Use self-teaching matching tests or simulate in-class testing with comprehensive multiple choice tests, or simply use Cram's true and false tests for quick review. Cram101.com even allows you to enter your in-class notes for an integrated studying format combining the textbook notes with your class notes.

Visit **www.Cram101.com**, click Sign Up at the top of the screen, and enter **DK73DW11731** in the promo code box on the registration screen. Your access to www.Cram101.com is discounted by 50% because you have purchased this book. Sign up and stop highlighting textbooks forever.

 ISBN(s): 9781614905554. PUBE-2.201155

Technical Communication
John M. Lannon, 11th

CONTENTS

1. PART I: Chapter 1 - Chapter 8 2
2. PART II: Chapter 9 - Chapter 16 42
3. PART III: Chapter 17 - Chapter 26 88

Audience	An Audience is a group of people who participate in a show or encounter a work of art, literature (in which they are called the 'reader'), theatre, music or academics in any medium. Audience members participate in different ways in different kinds of art; some events invite overt Audience participation and others allowing only modest clapping and criticism and reception. Media Audiences are studied by academics in media Audience studies.
Editing	Editing is the process of preparing language, images, sound, video condensation, organization, and other modifications in various media. A person who edits is called an editor. In a sense, the Editing process originates with the idea for the work itself and continues in the relationship between the author and the editor.
Letter	A letter is a written message from one party to another. The role of letters in communication has changed significantly since the 19th century. Historically, letters (in paper form) were the only reliable means of communication between two persons in different locations.
Organizations	Organizations · ALLARM · National Coalition on Racism in Sports and Media · STAR · Honor Indians Institute Articles · `Crimes Against Humanity` by Ward Churchill Parody · Turning the Tables on Racist Mascots · Alternative Indian Mascots

Simplification

In mathematical logic, Simplification is a valid argument and rule of inference which makes the inference that, if the conjunction A and B is true, then A is true, and B is true.

In formal language:

or

The argument has one premise, namely a conjunction, and one often uses Simplification in longer arguments to derive one of the conjuncts.

An example in English:

It`s raining and it`s pouring.

Supplement

A supplement is a publication that has a role secondary to that of another preceding or concurrent publication.

A follow-on publication complements its predecessor, either by bringing it up-to-date (e.g. the Index Catalogue), or by otherwise enhancing the predecessor's coverage of a particular topic or subject matter, as in the Tosefta. Supplements are particularly popular and useful in gaming hobbies.

Writing	Writing is the representation of language in a textual medium through the use of a set of signs or symbols (known as a writing system). It is distinguished from illustration, such as cave drawing and painting, and non-symbolic preservation of language via non-textual media, such as magnetic tape audio. Writing most likely began as a consequence of political expansion in ancient cultures, which needed reliable means for transmitting information, maintaining financial accounts, keeping historical records, and similar activities.
Writing style	Writing style is the manner in which a writer chooses among different strategies to address an issue and an audience. A style reveals the writer's personality or voice, but it also shows how she or he sees the audience of the writing. The writing style reveals the choices the writer makes in syntactical structures, diction, and figures of thought.
Adjustment	Adjustment means regulating, adapting or settling in a variety of contexts: Adjustment has several meanings; many relate to insurance, contracts, or the resolution of disputes. In Engineering, Mathematics and Geodesy, Adjustment means the optimal parameter estimation of a mathematical model so as to best fit a data set. The most important method is the least squares Adjustment, found by Carl Friedrich Gauss.
Byte	The byte, is a unit of digital information in computing and telecommunications, that most commonly consists of eight bits. Historically, a byte was the number of bits used to encode a single character of text in a computer and it is for this reason the basic addressable element in many computer architectures. The size of the byte has historically been hardware dependent and no definitive standards exist that mandate the size.

Chapter 1. PART I: Chapter 1 - Chapter 8

Communication	Communication is a process of transferring information from one entity to another. Communication processes are sign-mediated interactions between at least two agents which share a repertoire of signs and semiotic rules. Communication is commonly defined as `the imparting or interchange of thoughts, opinions, or information by speech, writing, or signs`.
Recommendation	A Recommendation can be: · In technical contexts, a norm (philosophy) or a norm (sociology) · European Union Recommendation, in international law · W3C Recommendation, in Internet contexts · a letter of Recommendation in employment or academia
Technical communication	Technical communication is the process of conveying technical information through writing, speech, and other mediums to a specific audience. Information is usable if the intended audience can perform an action or make a decision based on it (Johnson-Sheehan 7). Technical communicators often work collaboratively to create products (deliverables) for various media, including paper, video, and the Internet.
Fact sheet	A Fact sheet is a presentation of data on any subject in a format emphasizing brevity, key points of interest or concern, a fairly minimalist design aesthetic, and a general desire to convey the most relevant information in the least amount of space. As such, it is a medium ideally suited to the post-modern information age, when `facts` (to the extent they exist) proliferate in ever-increasing numbers while the time available to review and comprehend them diminishes. Fact sheets attempt to solve this problem by condensing information into an easy-to-read, straightforward, portable, and modular form of knowledge.
New media	New media is a term meant to encompass the emergence of digital, computerized, often having characteristics of being manipulatable, networkable, dense, compressible, interactive and impartial. Some examples may be the Internet, websites, computer multimedia, computer games, CD-ROMS, and DVDs.
Proposals	Proposals is a play by Neil Simon. Incidental music was written by Stephen Flaherty.

	A nostalgic memory play, proposals recalls one idyllic afternoon in the summer of 1953, the last time the Hines clan gathers at its retreat in the Poconos.
Reference card	A Reference card or quick Reference card is a concise bundling of common syntactic rules and idioms of a particular formal language. It serves as an ad hoc memory aid for an experienced user. In spite of what the name Reference card suggest, the term only tends to be used in the narrow context of programming languages or markup languages.
Blog	A blog is a type of website or part of a website. Blogs are usually maintained by an individual with regular entries of commentary, descriptions of events, or other material such as graphics or video. Entries are commonly displayed in reverse-chronological order.
Copyright	Copyright is the set of exclusive rights granted to the author or creator of an original work, including the right to copy, distribute and adapt the work. These rights can be licensed, transferred and/or assigned. Copyright lasts for a certain time period after which the work is said to enter the public domain.
Information	Information as a concept has many meanings, from everyday usage to technical settings. The concept of Information is closely related to notions of constraint, communication, control, data, form, instruction, knowledge, meaning, mental stimulus, pattern, perception, and representation. In its most restricted technical meaning, Information is an ordered sequence of symbols.
Paper	The Chinese invention of paper and the advent of woodblock printing produced the world\`s first print culture. As the print scholar A. Hyatt Mayor noted, \`it was the Chinese who really discovered the means of communication that was to dominate until our age.\` Another scholar, the prominent American historian and critic Daniel J. Boorstin, also noted in his book The Discoverers, singled out Korea as the \`most developed nation in the field of printing.\` Woodblock printing was better suited to Chinese characters than movable type, which the Chinese also invented, but which did not replace woodblock printing. In China and Korea, the use of woodblock printing on paper and movable type preceded their use in Europe by several centuries.
Technical communicators	Technical communicators use a variety of tools to create usable information. Often referred to as a \`tool set\`, there is no defined list of tools that are used by all technical writers. However, broad categories of tools are used by most technical writers and communicators.

Skills	A skill is the learned capacity to carry out pre-determined results often with the minimum outlay of time, energy, or both. skills can often be divided into domain-general and domain-specific skills. For example, in the domain of work, some general skills would include time management, teamwork and leadership, self motivation and others, whereas domain-specific skills would be useful only for a certain job.
Information technology	Information technology , as defined by the Information technology Association of America (Information technologyAA), is `the study, design, development, implementation, support or management of computer-based information systems, particularly software applications and computer hardware.` Information technology deals with the use of electronic computers and computer software to convert, store, protect, process, transmit, and securely retrieve information. Today, the term Information technology has ballooned to encompass many aspects of computing and technology, and the term has become very recognizable. Information technology professionals perform a variety of duties that range from installing applications to designing complex computer networks and information databases.
Differences	This article outlines the differences between Malaysian English or more popularly Manglish, the form of street Malaysian English spoken by most Malaysians and British English, which for the purposes of this article is assumed to be the form of English spoken in south east England, used by the British Government and the BBC and widely understood in other parts of the United Kingdom. It is necessary to make a distinction between Manglish and the English spoken by Malaysians speaking so-called proper English. While there are still certain peculiarities in the latter , proper Malaysian English is merely a normal variation in the way English is spoken and does not deviate significantly from common English.
Recommendation	A Recommendation can be: · In technical contexts, a norm (philosophy) or a norm (sociology) · European Union Recommendation, in international law · W3C Recommendation, in Internet contexts · a letter of Recommendation in employment or academia

Proofreading	Proofreading traditionally means reading a proof copy of a text in order to detect and correct any errors. Modern Proofreading often requires reading copy at earlier stages as well. A proof copy is a version of a manuscript that has been typeset after copy-editing.
Punctuation	Punctuation marks are symbols that indicate the structure and organization of written language, as well as intonation and pauses to be observed when reading aloud. In written English, punctuation is vital to disambiguate the meaning of sentences. For example, "woman, without her man, is nothing" and "woman: without her, man is nothing" have greatly different meanings, as do "eats shoots and leaves" and "eats, shoots and leaves".
Sentence	In the field of linguistics, a sentence is an expression in natural language, and often defined to indicate a grammatical unit consisting of one or more words that generally bear minimal syntactic relation to the words that precede or follow it. A sentence can include words grouped meaningfully to express a statement, question, exclamation, request, command or suggestion. As with all language expressions, sentences may contain both function and content words, and contain properties distinct to natural language, such as characteristic intonation and timing patterns.
Typographical error	A typographical error is a mistake made in, originally, the manual type-setting (typography) of printed material, or more recently, the typing process. The term includes errors due to mechanical failure or slips of the hand or finger, but usually excludes errors of ignorance, such as spelling errors. Before the arrival of printing, the "copyist's mistake" or "scribal error" was the equivalent for manuscripts.

Apostrophe	The apostrophe is a punctuation mark, and sometimes a diacritic mark, in languages that use the Latin alphabet or certain other alphabets. In English, it serves three purposes: • The omission of one or more letters (as in the contraction of do not to don't). • The marking of possessive case (as in the cat's whiskers). • The marking as plural of written items that are not words established in English orthography (as in P's and Q's, the late 1950's). (This is considered incorrect by some.
Comma	The comma is a punctuation mark. It has the same shape as an apostrophe or single closing quotation mark in many typefaces, but it differs from them in being placed on the baseline of the text. Some typefaces render it as a small line, slightly curved or straight, or with the appearance of a small, filled-in number 9.
Desktop publishing	Desktop publishing (also known as DTP) combines a personal computer and WYSIWYG page layout software to create publication documents on a computer for either large scale publishing or small scale local multifunction peripheral output and distribution. The term 'Desktop publishing' is commonly used to describe page layout skills. However, the skills and software are not limited to paper and book publishing.
Employee	Employment is a contract between two parties, one being the employer and the other being the employee. An employee may be defined as: `A person in the service of another under any contract of hire, express or implied, oral or written, where the employer has the power or right to control and direct the employee in the material details of how the work is to be performed.` Black`s Law Dictionary page 471 (5th ed. 1979). In a commercial setting, the employer conceives of a productive activity, generally with the intention of generating a profit, and the employee contributes labour to the enterprise, usually in return for payment of wages.
Setting	In fiction, setting includes the time, location, and everything in which a story takes place, and initiates the main backdrop and mood for a story. Setting has been referred to as story world or milieu to include a context (especially society) beyond the immediate surroundings of the story. Elements of setting may include culture, historical period, geography, and hour.

Usability	Usability is the ease of use and learnability of a human-made object. The object of use can be a software application, website, book, tool, machine, process, or anything a human interacts with. A usability study may be conducted as a primary job function by a usability analyst or as a secondary job function by designers, technical writers, marketing personnel, and others.
Outline	An outline is a list of the main features of a given topic, often used as a rough draft or summary of the content of a document. A hierarchical outline is a list arranged to show hierarchical relationships, and is a type of tree structure. Writers of fiction and creative nonfiction, such as Jon Franklin, may use outlines to establish plot sequence, character development and dramatic flow of a story, sometimes in conjunction with freewriting.
Interpretation	An interpretation is an assignment of meaning to the symbols of a language. Many formal languages used in mathematics, logic, and theoretical computer science are defined in solely syntactic terms, and as such do not have any meaning until they are given some interpretation. The general study of interpretations of formal languages is called formal semantics.
Revision	Revision is the stage in the writing process where the author reviews, alters, and amends her or his message, according to what has been written in the draft. Revision follows drafting and precedes editing. Drafting and revising often form a loop as a work moves back and forth between the two stages.
Internet	The Internet is a global system of interconnected computer networks that use the standardized Internet Protocol Suite (TCP/IP) to serve billions of users worldwide. It is a network of networks that consists of millions of private and public, academic, business, and government networks of local to global scope that are linked by copper wires, fiber-optic cables, wireless connections, and other technologies. The Internet carries a vast array of information resources and services, most notably the inter-linked hypertext documents of the World Wide Web (WWW) and the infrastructure to support electronic mail.
On Writing	On Writing: A Memoir of the Craft is an autobiography and writing guide by Stephen King, published in 2000. It is a book about the prolific author's experiences as a writer. In 2008, Entertainment Weekly listed On Writing 21st on their list of The New Classics: Books - The 100 best reads from 1983 to 2008, making it King's only entry. On July 13, 2010, Scribner published a 10th Anniversary edition of the book (ISBN 978-1-4391-9363-1).

Argument	An argument in literature is a brief summary, often in prose, of a poem or section of a poem or other work. It is often appended to the beginning of each chapter, book, or canto. They were common during the Renaissance as a way to orient a reader within a large work.
Prediction	A prediction is a statement or claim that a particular event will occur in the future in more certain terms than a forecast. The etymology of the word is Latin . Howard H. Stevenson writes: \`prediction is at least two things: Important and hard.\` Important, because we have to act, and hard because we have to realize the future we want, and what is the best way to get there.
Authority	The authority of a text is its reliability as a witness to the author's intentions. These intentions could be initial, medial or final, but intentionalist editors (most notably represented by Fredson Bowers and G. Thomas Tanselle editing school) generally attempt to retrieve final authorial intentions. The concept is of particular importance for textual critics, whether they believe that authorial intention is recoverable, or whether they think that this recovery is impossible.
Statement	Statement may refer to: • A kind of expression in language (linguistics) • Statement declarative sentence that is either true or false • Statement the smallest standalone element of an imperative programming language • "Statement" (song), 2008 song by the Japanese band Boris • Statements (album), 1962 album by jazz vibraphonist Milt Jackson • Financial statements, formal records of the financial activities of a business, person, or other entity • Mathematical statement, a statement in logic and mathematics • Political statement, any act or nonverbal form of communication that is intended to influence a decision to be made for or by a group • Press statement, written or recorded communication directed at members of the news media • Statement of Special Educational Needs, outlining specific provision needed for a child in England • a paper size also known as organizer L or half letter .
Production	Production may be: In Economics:

· Production, costs, and pricing, the act of making products (goods and services)

· Production, the act of manufacturing goods

· Production as statistic, gross domestic product

· A Production line
In Entertainment:

· Production, phase of filmmaking

· Production, category of illusory magic trick

· Production, film distributor of a company

· Production, theatrical performance or theatre producer
Abstract systems:

· Production (computer science), formal-grammar concept

· Production system, programming method .

Double bind

A Double bind is a dilemma in communication in which an individual receives two or more conflicting messages, with one message negating the other. This creates a situation in which a successful response to one message implicates a failed response to the other, so that the person will be automatically wrong regardless of response. The person can neither comment on the conflict, nor resolve it, nor opt out of the situation.

Group dynamics	Group dynamics is the study of groups, and also a general term for group processes. Relevant to the fields of psychology, sociology, and communication studies, a group is two or more individuals who are connected to each other by social relationships. Because they interact and influence each other, groups develop a number of dynamic processes that separate them from a random collection of individuals.
Knowledge	Knowledge is defined by the Oxford English Dictionary as expertise, and skills acquired by a person through experience or education; the theoretical or practical understanding of a subject, (ii) what is known in a particular field or in total; facts and information or (iii) awareness or familiarity gained by experience of a fact or situation. Philosophical debates in general start with Plato`s formulation of knowledge as `justified true belief`. There is however no single agreed definition of knowledge presently, nor any prospect of one, and there remain numerous competing theories.
Conflict	Conflict is a necessary part of our lives and is a very real issue. Sir Arthur Thomas Quiller-Couch, literary critic and author, was first to classify plots as seven basic conflicts: Man against Man, Man against Nature, Man against Himself, Man against God, Man against Society, Man caught in the Middle, Man ' Woman. This has inspired a variety of similar lists, as follows: Common examples of conflict Character vs. Self Character vs. Self is when the main character in the story has a problem with him or herself.
Intellectual property	Intellectual property is a number of distinct types of legal monopolies over creations of the mind, both artistic and commercial, and the corresponding fields of law. Under Intellectual property law, owners are granted certain exclusive rights to a variety of intangible assets, such as musical, literary, and artistic works; ideas, discoveries and inventions; and words, phrases, symbols, and designs. Common types of Intellectual property include copyrights, trademarks, patents, industrial design rights and trade secrets in some jurisdictions.

Employee

Employment is a contract between two parties, one being the employer and the other being the employee. An employee may be defined as: `A person in the service of another under any contract of hire, express or implied, oral or written, where the employer has the power or right to control and direct the employee in the material details of how the work is to be performed.` Black`s Law Dictionary page 471 (5th ed. 1979).

In a commercial setting, the employer conceives of a productive activity, generally with the intention of generating a profit, and the employee contributes labour to the enterprise, usually in return for payment of wages.

False advertising

False advertising or deceptive advertising is the use of false or misleading statements in advertising. As advertising has the potential to persuade people into commercial transactions that they might otherwise avoid, many governments around the world use regulations to control false, deceptive or misleading advertising. Truth in labeling refers to essentially the same concept, that customers have the right to know what they are buying, and that all necessary information should be on the label.

Source

A source or sender is one of the basic concepts of communication and information processing. sources are objects which encode message data and transmit the information, via a channel, to one or more observers (or receivers).
In the strictest sense of the word, particularly in information theory, a source is a process that generates message data that one would like to communicate, or reproduce as exactly as possible somewhere else in space or time.

Documentation

Documentation is a general term for a multiplicity of documents in a chosen mix of media and with a certain collection. Purpose of documentation is the use to support a tool or a process.

Classical documentation is a set of documents printed on paper.

Plagiarism	Plagiarism, as defined in the 1995 Random House Compact Unabridged Dictionary, is the `use or close imitation of the language and thoughts of another author and the representation of them as one`s own original work.` Within academia, Plagiarism by students, professors, or researchers is considered academic dishonesty or academic fraud, and offenders are subject to academic censure, up to and including expulsion. In journalism, Plagiarism is considered a breach of journalistic ethics, and reporters caught plagiarizing typically face disciplinary measures ranging from suspension to termination of employment. Some individuals caught plagiarizing in academic or journalistic contexts claim that they plagiarized unintentionally, by failing to include quotations or give the appropriate citation.
Meeting	In a Meeting, two or more people come together for the purpose of discussing a (usually) predetermined topic such as business or community event planning, often in a formal setting. In addition to coming together physically (in real life, face to face), communication lines and equipment can also be set up to have a discussion between people at different locations, e.g. a conference call or an e-Meeting. In organizations, Meetings are an important vehicle for personal contact.
Conducting	Conducting is the act of directing a musical performance by way of visible gestures. Orchestras, choirs, concert bands and other musical ensembles often have conductors. The principal conductor of an orchestra or opera company is sometimes referred to as a music director or chief conductor, or by the German word, Kapellmeister.
Group conflicts	Group conflicts is where social behaviour causes groups of individuals to conflict with each other. It can also refer to a conflict within these groups. This conflict is often caused by differences in social norms, values, and religion.
Brainstorming	Brainstorming is a group creativity technique designed to generate a large number of ideas for the solution of a problem. In 1953 the method was popularized by Alex Faickney Osborn in a book called Applied Imagination. Osborn proposed that groups could double their creative output with Brainstorming.
Brainwriting	6-3-5 Brainwriting is a group creativity technique used in marketing, advertising, design, writing and product development originally developed by Professor Bernd Rohrbach in 1968.

	Based on the concept of Brainstorming, the aim of 6-3-5 Brainwriting is to generate 108 new ideas in half an hour. In a similar way to brainstorming, it is not the quality of ideas that matters but the quantity.
Esophageal speech	Esophageal speech (or voice) is an alternate method for speech production without the oscillation in the vocal folds. The sound is produced by releasing gases from or through the esophagus. In Esophageal speech it is thus the esophagus that oscillates in contrast to normal (laryngeal) speech where the vocal folds oscillate.
Digital	A digital system is a data technology that uses discrete (discontinuous) values. By contrast, non-digital (or analog) systems use a continuous range of values to represent information. Although digital representations are discrete, the information represented can be either discrete, such as numbers, letters or icons, or continuous, such as sounds, images, and other measurements of continuous systems.
Instant messaging	Instant messaging is a form of real-time communication between two or more people based on typed text. The text is conveyed via devices connected over a network such as the Internet. Instant messaging is a collection of technologies that create the possibility of real-time text-based communication between two or more participants over the internet or some form of internal network/intranet.
Videoconferencing	A videoconference or video conference (also known as a videoteleconference) is a set of interactive telecommunication technologies which allow two or more locations to interact via two-way video and audio transmissions simultaneously. It has also been called `visual collaboration` and is a type of groupware.
	videoconferencing differs from a videophone calls in that its designed to serve a conference rather than individuals.
Collaboration	Collaboration is a recursive process where two or more people or organizations work together in an intersection of common goals -- for example, an intellectual endeavor that is creative in nature--by sharing knowledge, learning and building consensus. Most Collaboration requires leadership, although the form of leadership can be social leadership within a decentralized and egalitarian group. In particular, teams that work collaboratively can obtain greater resources, recognition and reward when facing competition for finite resources.Collaboration is also present in opposing goals exhibiting the notion of adversarial Collaboration, though this is not a common case for using the term.

Chunking	Chunking is a method of presenting information which splits concepts into small pieces or "chunks" of information to make reading and understanding faster and easier. Chunking is especially useful for material presented on the web because readers tend to scan for specific information on a web page rather than read the page sequentially. Chunked content usually contains: • bulleted lists • short subheadings • short sentences with one or two ideas per sentence • short paragraphs, even one-sentence paragraphs • easily scannable text, with bolding of key phrases • inline graphics to guide the eyes or illustrate points which would normally require more words Advantages of chunking • Chunking helps technical communicators or marketers convey information more efficiently • Chunking helps readers find what they are looking for quickly • Chunking allows material to be presented consistently from page to page, so users can apply previous knowledge of page layout and navigation and focus on the content rather than the presentation Disadvantages of chunking Monotonous or repetitive layout is seldom memorable.
Credit	In general, the term credit in the artistic or intellectual sense refers to an acknowledgement of those who contributed to a work, whether through ideas or in a more direct sense. Credit in the arts

	In the creative arts, credits are an acknowledgement of those who participated in the production. They are often shown at the end of movies and on CD jackets.
Collaborative writing	The term collaborative writing refers to projects where written works are created by multiple people together (collaboratively) rather than individually. Some projects are overseen by an editor or editorial team, but many grow without any oversight. Practical approaches In a true collaborative environment, each contributor has an almost equal ability to add, edit, and remove text.
Writing process	The Writing process is both a key concept in the teaching of writing and an important research concept in the field of composition studies. Research on the writing process focuses on how writers draft, revise, and edit texts. Composing process research was pioneered by scholars such as Janet Emig in "The Composing Processes of Twelfth Graders" (1971), Sondra Perl in "The Composing Processes of Unskilled College Writers (1979), and Linda Flower and John R. Hayes in "A Cognitive Process Theory of Writing" (1981).
Balance	Balance or impartiality, is , Lord John Reith, the BBC's founding General Manager and later Chairman, promoted a concept called Balance. In practise Balance means ensuring that statements by those challenging the establishment are Balanced with statements of those whom they are criticising, though not necessarily the other way round.
Google	Google Inc. is an American public corporation, earning revenue from advertising related to its Internet search, e-mail, online mapping, office productivity, social networking, and video sharing services as well as selling advertising-free versions of the same technologies. Google has also developed an open source web browser and a mobile operating system.

Abstract	An abstract is a brief summary of a research article, thesis, review, conference proceeding or any in-depth analysis of a particular subject or discipline, and is often used to help the reader quickly ascertain the paper's purpose. When used, an abstract always appears at the beginning of a manuscript, acting as the point-of-entry for any given scientific paper or patent application. Abstracting and indexing services for various academic disciplines are aimed at compiling a body of literature for that particular subject.
Magazine	Magazines, periodicals, glossies or serials are publications, generally published on a regular schedule, containing a variety of articles, generally financed by advertising, by a purchase price, by pre-paid magazine subscriptions, bookstores or other vendors; or through free distribution at selected pick up locations. The various elements that contribute to the production of magazines vary wildly.
Magazine	Magazines, periodicals, glossies or serials are publications, generally published on a regular schedule, containing a variety of articles, generally financed by advertising, by a purchase price, by pre-paid magazine subscriptions, bookstores or other vendors; or through free distribution at selected pick up locations. The various elements that contribute to the production of magazines vary wildly.
Community	In biological terms, a Community is a group of interacting organisms sharing an environment. In human communities, intent, belief, resources, preferences, needs, risks, and a number of other conditions may be present and common, affecting the identity of the participants and their degree of cohesiveness. In sociology, the concept of Community has caused infinite debate, and sociologists are yet to reach agreement on a definition of the term.
Double bind	A Double bind is a dilemma in communication in which an individual receives two or more conflicting messages, with one message negating the other. This creates a situation in which a successful response to one message implicates a failed response to the other, so that the person will be automatically wrong regardless of response. The person can neither comment on the conflict, nor resolve it, nor opt out of the situation.
Extranet	An Extranet is a private network that uses Internet protocols, network connectivity. An Extranet can be viewed as part of a company`s intranet that is extended to users outside the company, usually via the Internet. It has also been described as a `state of mind` in which the Internet is perceived as a way to do business with a selected set of other companies (business-to-business, B2B), in isolation from all other Internet users.

Disc	Disc and disk are the two alternative spellings of the descriptive word for things of a generally thin and circular geometry. These variations are due to the way in which the words originated. The Discussion here somewhat focuses on how the word applies to data storage media.
Guide	A Guide is a person who leads anyone through unknown or unmapped country. This includes a Guide of the real world (such as someone who conducts travellers and tourists through a place of interest), as well as a person who leads someone to more abstract places (such as to knowledge or wisdom). The word 'Guide' was incorporated into English via Old French 'Guider' which meant 'to Guide, lead, conduct' it was originally taken by Old French from Frankish '*witan' meaning 'show the way' from Proto-Germanic '*wit-' meaning 'to know' .
Literature	Literature is the art of written works, and is not bound to published sources (although, under circumstances they can be exempt). Literally translated, the word literature means "acquaintance with letters" (as in the "arts and letters"). The two most basic written literary categories include fiction and non fiction.
Article	An article is a written work published in a print or Internet medium. It may be for the purpose of propagating the news, research results, academic analysis or debatee. daily newspapers) or on a specific topic (i.e. political or trade news magazines, club newsletters, or technology news websites).
Newspaper	A Newspaper is a publication containing news, information, and advertising. General-interest Newspapers often feature articles on political events, crime, business, art/entertainment, society and sports. Most traditional papers also feature an editorial page containing columns that express the personal opinions of writers.
Technical report	A technical report is a document that describes the process, progress, or results of technical or scientific research or the state of a technical or scientific research problem. It might also include recommendations and conclusions of the research. Unlike other scientific literature, such as scientific journals and the proceedings of some academic conferences, technical reports rarely undergo comprehensive independent peer review before publication.
Index	The symbol ? is a punctuation mark, called an index, manicule or fist. Though rare today, this symbol was in common use between the 12th and 18th centuries in the margins of books, and was formerly included in lists of standard punctuation marks. Its typical use is as a bullet-like symbol to direct the reader's attention to important text, having roughly the same meaning as the word "attention" or "note".
Conference	A conference is a meeting of people that 'confer' about a topic.

· Academic conference, in science and academia, a formal event where researchers present results, workshops, and other activities.

· Business conference, organized to discuss business-related matters best effected there.

Publication

The word Publication means the act of publishing, and it also means any writing of which copies are published, and any website. Among Publications are books, and periodicals, the latter including magazines, scholarly journals, and newspapers.
`Publication` is a technical term in legal contexts and especially important in copyright legislation.

Microform

Microforms are any forms, either films or paper, containing microreproductions of documents for transmission, storage, reading, and printing. Microform images are commonly reduced about 25 times from the original document size. For special purposes, greater optical reductions may be used.

Internet

The Internet is a global system of interconnected computer networks that use the standardized Internet Protocol Suite (TCP/IP) to serve billions of users worldwide. It is a network of networks that consists of millions of private and public, academic, business, and government networks of local to global scope that are linked by copper wires, fiber-optic cables, wireless connections, and other technologies. The Internet carries a vast array of information resources and services, most notably the inter-linked hypertext documents of the World Wide Web (WWW) and the infrastructure to support electronic mail.

Loaded question

Loaded question presupposition, `trick question`, is an informal fallacy or logical fallacy. It is committed when someone asks a question that presupposes something that has not been proven or accepted by all the people involved. This fallacy is often used rhetorically, so that the question limits direct replies to those that serve the questioner`s agenda.

Source

A source or sender is one of the basic concepts of communication and information processing. sources are objects which encode message data and transmit the information, via a channel, to one or more observers (or receivers).
In the strictest sense of the word, particularly in information theory, a source is a process that generates message data that one would like to communicate, or reproduce as exactly as possible somewhere else in space or time.

Double bind

A Double bind is a dilemma in communication in which an individual receives two or more conflicting messages, with one message negating the other. This creates a situation in which a successful response to one message implicates a failed response to the other, so that the person will be automatically wrong regardless of response. The person can neither comment on the conflict, nor resolve it, nor opt out of the situation.

Closed-ended question

A closed-ended question is a form of question which can normally be answered using a simple "yes" or "no", a specific simple piece of information, or a selection from multiple choices.

Examples include:

- Question: Do you know your weight?

Answer: Yes.

- Question: What is your weight?

	Answer: 167 lbs.
Letter	A letter is a written message from one party to another. The role of letters in communication has changed significantly since the 19th century. Historically, letters (in paper form) were the only reliable means of communication between two persons in different locations.
Information	Information as a concept has many meanings, from everyday usage to technical settings. The concept of Information is closely related to notions of constraint, communication, control, data, form, instruction, knowledge, meaning, mental stimulus, pattern, perception, and representation. In its most restricted technical meaning, Information is an ordered sequence of symbols.
Literature	Literature is the art of written works, and is not bound to published sources (although, under circumstances they can be exempt). Literally translated, the word literature means "acquaintance with letters" (as in the "arts and letters"). The two most basic written literary categories include fiction and non fiction.
Publication	The word Publication means the act of publishing, and it also means any writing of which copies are published, and any website. Among Publications are books, and periodicals, the latter including magazines, scholarly journals, and newspapers. `Publication` is a technical term in legal contexts and especially important in copyright legislation.
Chunking	Chunking is a method of presenting information which splits concepts into small pieces or "chunks" of information to make reading and understanding faster and easier. Chunking is especially useful for material presented on the web because readers tend to scan for specific information on a web page rather than read the page sequentially.

Chunked content usually contains:

- bulleted lists
- short subheadings
- short sentences with one or two ideas per sentence
- short paragraphs, even one-sentence paragraphs
- easily scannable text, with bolding of key phrases
- inline graphics to guide the eyes or illustrate points which would normally require more words

Advantages of chunking

- Chunking helps technical communicators or marketers convey information more efficiently
- Chunking helps readers find what they are looking for quickly
- Chunking allows material to be presented consistently from page to page, so users can apply previous knowledge of page layout and navigation and focus on the content rather than the presentation

Disadvantages of chunking

Monotonous or repetitive layout is seldom memorable.

Interpretation

An interpretation is an assignment of meaning to the symbols of a language. Many formal languages used in mathematics, logic, and theoretical computer science are defined in solely syntactic terms, and as such do not have any meaning until they are given some interpretation. The general study of interpretations of formal languages is called formal semantics.

Blog

A blog is a type of website or part of a website. Blogs are usually maintained by an individual with regular entries of commentary, descriptions of events, or other material such as graphics or video. Entries are commonly displayed in reverse-chronological order.

Double bind

A Double bind is a dilemma in communication in which an individual receives two or more conflicting messages, with one message negating the other. This creates a situation in which a successful response to one message implicates a failed response to the other, so that the person will be automatically wrong regardless of response. The person can neither comment on the conflict, nor resolve it, nor opt out of the situation.

Proofs	In mathematics, there is a variety of spurious proofs of obvious contradictions. Although the proofs are flawed, the errors, usually by design, are comparatively subtle. These fallacies are normally regarded as mere curiosities, but can be used to show the importance of rigor in mathematics.
Audience	An Audience is a group of people who participate in a show or encounter a work of art, literature (in which they are called the 'reader'), theatre, music or academics in any medium. Audience members participate in different ways in different kinds of art; some events invite overt Audience participation and others allowing only modest clapping and criticism and reception. Media Audiences are studied by academics in media Audience studies.
Argument	An argument in literature is a brief summary, often in prose, of a poem or section of a poem or other work. It is often appended to the beginning of each chapter, book, or canto. They were common during the Renaissance as a way to orient a reader within a large work.
Mode	In literature, a mode is an employed method or approach, identifiable within a written work. As descriptive terms, form and genre are often used inaccurately instead of mode; for example, the pastoral mode is often mistakenly identified as a genre. The Writers Web site feature, A List of Important Literary Terms, defines mode thus: An unspecific critical term usually identifying a broad, but identifiable literary method, mood, or manner, that is not tied exclusively to a particular form or genre.
Data mining	Data mining is the process of extracting patterns from data. Data mining is becoming an increasingly important tool to transform these data into information. It is commonly used in a wide range of profiling practices, such as marketing, surveillance, fraud detection and scientific discovery.
Margin of error	The Margin of error is a statistic expressing the amount of random sampling error in a survey\`s results. The larger the Margin of error, the less faith one should have that the poll\`s reported results are close to the \`true\` figures; that is, the figures for the whole population. The Margin of error is usually defined as the \`radius\` (or half the width) of a confidence interval for a particular statistic from a survey.

Terminology	Terminology is the study of terms and their use. Terms are words and compound words that are used in specific contexts. Not to be confused with "terms" in colloquial usages, the shortened form of technical terms (or terms of art) which are defined within a discipline or speciality field.
Bias	Bias is a term used to describe a tendency or preference towards a particular perspective, ideology or result, when the tendency interferes with the ability to be impartial, unprejudiced, or objective.. In other words, bias is generally seen as a 'one-sided' perspective. The term biased refers to a person or group who is judged to exhibit bias.
Abstract	An abstract is a brief summary of a research article, thesis, review, conference proceeding or any in-depth analysis of a particular subject or discipline, and is often used to help the reader quickly ascertain the paper's purpose. When used, an abstract always appears at the beginning of a manuscript, acting as the point-of-entry for any given scientific paper or patent application. Abstracting and indexing services for various academic disciplines are aimed at compiling a body of literature for that particular subject.
Case study	A case study is one of several ways of doing research whether it is social science related or even socially related. It is an in-depth investigation/study of a single individual, group, incident, or community. Other ways include experiments, surveys, or analysis of archival information .
Executive summary	Executive summary is a term used in business for a short document that summarizes a longer report, proposal or group of related reports in such a way that readers can rapidly become acquainted with a large body of material without having to read it all. It will usually contain a brief statement of the problem or proposal covered in the major document(s), background information, concise analysis and main conclusions. It is intended as an aid to decision making by business managers.
Description	Description is one of four rhetorical modes (also known as modes of discourse), along with exposition, argumentation, and narration. Each of the rhetorical modes is present in a variety of forms and each has its own purpose and conventions. Description is also the fiction-writing mode for transmitting a mental image of the particulars of a story.
Editing	Editing is the process of preparing language, images, sound, video condensation, organization, and other modifications in various media. A person who edits is called an editor. In a sense, the Editing process originates with the idea for the work itself and continues in the relationship between the author and the editor.
Organizations	Organizations

· ALLARM

· National Coalition on Racism in Sports and Media

· STAR

· Honor Indians Institute

Articles

· `Crimes Against Humanity` by Ward Churchill

Parody

· Turning the Tables on Racist Mascots

· Alternative Indian Mascots

Usability

Usability is the ease of use and learnability of a human-made object. The object of use can be a software application, website, book, tool, machine, process, or anything a human interacts with. A usability study may be conducted as a primary job function by a usability analyst or as a secondary job function by designers, technical writers, marketing personnel, and others.

Writing

Writing is the representation of language in a textual medium through the use of a set of signs or symbols (known as a writing system). It is distinguished from illustration, such as cave drawing and painting, and non-symbolic preservation of language via non-textual media, such as magnetic tape audio.

Writing most likely began as a consequence of political expansion in ancient cultures, which needed reliable means for transmitting information, maintaining financial accounts, keeping historical records, and similar activities.

Writing style

Writing style is the manner in which a writer chooses among different strategies to address an issue and an audience. A style reveals the writer's personality or voice, but it also shows how she or he sees the audience of the writing. The writing style reveals the choices the writer makes in syntactical structures, diction, and figures of thought.

Outline

An outline is a list of the main features of a given topic, often used as a rough draft or summary of the content of a document. A hierarchical outline is a list arranged to show hierarchical relationships, and is a type of tree structure.

Writers of fiction and creative nonfiction, such as Jon Franklin, may use outlines to establish plot sequence, character development and dramatic flow of a story, sometimes in conjunction with freewriting.

Outline

An outline is a list of the main features of a given topic, often used as a rough draft or summary of the content of a document. A hierarchical outline is a list arranged to show hierarchical relationships, and is a type of tree structure.

Writers of fiction and creative nonfiction, such as Jon Franklin, may use outlines to establish plot sequence, character development and dramatic flow of a story, sometimes in conjunction with freewriting.

Introduction

An introduction is a beginning section which states the purpose and goals of the following writing. The introduction is usually interesting and it intrigues the reader and causes him or her to want to read on. The sentence in which the introduction begins can be a question or just a statement.

Alphanumeric

Alphanumeric is a combination of alphabetic and numeric , and is used to describe the collection of Latin letters and Arabic digits used by much of western society. There are either 36 or 62 (case-sensitive) Alphanumeric characters. The Alphanumeric character set consists of the numbers 0 to 9 and letters A to Z.

In computing terminology, a character stored in Alphanumeric form is considerably smaller than storing an 8-bit ASCII character, as each character is only 6 bits in length.

Notations

This page is an overview of the common notations used when working with vectors, which may be spatial or more abstract members of vector spaces.

The common typographic convention for representing a vector is upright boldface type, as in v for a vector named `v`. In handwriting, where boldface type is either unavailable or unwieldy, vectors are often represented with right-pointing arrows or harpoons above their names, as in .

Sentence

In the field of linguistics, a sentence is an expression in natural language, and often defined to indicate a grammatical unit consisting of one or more words that generally bear minimal syntactic relation to the words that precede or follow it. A sentence can include words grouped meaningfully to express a statement, question, exclamation, request, command or suggestion.

As with all language expressions, sentences may contain both function and content words, and contain properties distinct to natural language, such as characteristic intonation and timing patterns.

Statement

Statement may refer to:

- A kind of expression in language (linguistics)
- Statement declarative sentence that is either true or false
- Statement the smallest standalone element of an imperative programming language
- "Statement" (song), 2008 song by the Japanese band Boris
- Statements (album), 1962 album by jazz vibraphonist Milt Jackson
- Financial statements, formal records of the financial activities of a business, person, or other entity
- Mathematical statement, a statement in logic and mathematics
- Political statement, any act or nonverbal form of communication that is intended to influence a decision to be made for or by a group
- Press statement, written or recorded communication directed at members of the news media
- Statement of Special Educational Needs, outlining specific provision needed for a child in England
- a paper size also known as organizer L or half letter

.

Modifier	In grammar, a modifier is an optional element in phrase structure or clause structure; the removal of the modifier typically doesn`t affect the grammaticality of the construction. modifiers can be a word, a phrase or an entire clause. Semantically, modifiers describe and provide more accurate definitional meaning for another element.
Pronoun	In linguistics and grammar, a Pronoun is a pro-form that substitutes for a noun with or without a determiner, such as you and they in English. The replaced phrase is called the antecedent of the Pronoun. For example, consider the sentence `Lisa gave the coat to Phil.` All three nouns in the sentence can be replaced by Pronouns: `She gave it to him.` If the coat, Lisa, and Phil have been previously mentioned, the listener can deduce what the Pronouns she, it and him refer to and therefore understand the meaning of the sentence; however, if the sentence `She gave it to him.` is the first presentation of the idea, none of the Pronouns have antecedents, and each Pronoun is therefore ambiguous.
Esophageal speech	Esophageal speech (or voice) is an alternate method for speech production without the oscillation in the vocal folds. The sound is produced by releasing gases from or through the esophagus. In Esophageal speech it is thus the esophagus that oscillates in contrast to normal (laryngeal) speech where the vocal folds oscillate.
Noun	In linguistics, a Noun is a member of a large, open lexical category whose members can occur as the main word in the subject of a clause, the object of a verb, or the object of a preposition. Lexical categories are defined in terms of how their members combine with other kinds of expressions. The syntactic rules for Nouns differ from language to language.
Passive voice	Passive voice is a grammatical voice common in many of the world`s languages. Passive is used in a clause whose subject expresses the theme or patient of the main verb. That is, the subject undergoes an action or has its state changed.
Sentence	In the field of linguistics, a sentence is an expression in natural language, and often defined to indicate a grammatical unit consisting of one or more words that generally bear minimal syntactic relation to the words that precede or follow it. A sentence can include words grouped meaningfully to express a statement, question, exclamation, request, command or suggestion.

	As with all language expressions, sentences may contain both function and content words, and contain properties distinct to natural language, such as characteristic intonation and timing patterns.
Wordiness	Wordiness is the use of more words than necessary. As it makes a text slightly harder to read, it is usually discouraged.
Verb	In syntax, a Verb, deriving from the Latin Verbum meaning word, is a word that conveys action (bring, read, walk, run, murder), or a state of being (exist, stand). In most languages, Verbs are inflected (modified in form) to encode tense, aspect, mood and voice. A Verb may also agree with the person, gender, and/or number of some of its arguments, such as its subject, or object.
Preposition	In grammar, a preposition is a part of speech that introduces a prepositional phrase. For example, in the sentence \`The cat sleeps on the sofa\`, the word \`on\` is a preposition, introducing the prepositional phrase \`on the sofa\`. In English, the most used prepositions are \`of\`, \`to\`, \`in\`, \`for\`, \`with\` and \`on\`.
Nominalization	Nominalization in linguistics, (1) is the use of a verb or an adjective or an adverb as a noun with or without morphological transformation, so that the word can now act as the head of a noun phrase, and (2) is the process of producing a noun by adding derivational affixes to another word classes (e.g. legalize-legalization). Nominalization happens in languages around the world. Some languages allow verbs to be used as nouns, while others require some form of morphological transformation.
Fluency	Fluency is the property of a person or of a system that delivers information quickly and with expertise. fluency is a speech language pathology term that means the smoothness or flow with which sounds, syllables, words and phrases are joined together when speaking quickly.
Punctuation	Punctuation marks are symbols that indicate the structure and organization of written language, as well as intonation and pauses to be observed when reading aloud.

	In written English, punctuation is vital to disambiguate the meaning of sentences. For example, "woman, without her man, is nothing" and "woman: without her, man is nothing" have greatly different meanings, as do "eats shoots and leaves" and "eats, shoots and leaves".
Independent	The National Literary Awards defines Independent or \`Indie\` Literature as \`books published outside of mainstream publishing.\` Such are books rarely recognized and hard to pin down, but some examples include \`Returning Home\` by Marcus Blake and \`Hope .
Independent clause	An Independent clause is a clause that can stand by itself, also known as a simple sentence. An Independent clause contains a subject and a predicate; it makes sense by itself. Multiple Independent clauses can be joined by using a semicolon or a comma plus a coordinating conjunction (for, and, nor, but, or, yet, so). · I love penguins. (simple sentence) · I drive a bus. (simple sentence) · I am a doctor, and my wife is a lawyer. · I want to be a nurse, but I need to receive my science degree. (I need to receive my science degree) · Go to the store, and get me a carton of milk.
Jargon	Jargon is terminology which is especially defined in relationship to a specific activity, profession, group, or event. The philosophe Condillac observed in 1782 that "Every science requires a special language because every science has its own ideas." As a rationalist member of the Enlightenment he continued, "It seems that one ought to begin by composing this language, but people begin by speaking and writing and the language remains to be composed."

In other words, the term covers the language used by people who work in a particular area or who have a common interest. Much like slang, it can develop as a kind of short-hand, to express ideas that are frequently discussed between members of a group, though it can also be developed deliberately using chosen terms.

Acronyms

Acronyms and initialisms are abbreviations that are formed using the initial components in a phrase or name. These components may be individual letters (as in CEO) or parts of words (as in Benelux). There is no universal agreement on the precise definition of the various terms , nor on written usage .

Classroom

A classroom is a room in which teaching or learning activities can take place. classrooms are found in educational institutions of all kinds, including public and private schools, corporations, and religious and humanitarian organizations. The classroom attempts to provide a safe space where learning can take place uninterrupted by other distractions.

Modem

Modem is a device that modulates an analog carrier signal to encode digital information, and also demodulates such a carrier signal to decode the transmitted information. The goal is to produce a signal that can be transmitted easily and decoded to reproduce the original digital data. Modems can be used over any means of transmitting analog signals, from driven diodes to radio.

Euphemism

A Euphemism is a substitution of an agreeable or less offensive expression in place of one that may offend or suggest something unpleasant to the listener, to make it less troublesome for the speaker. It also may be a substitution of a description of something or someone rather than the name, to avoid revealing secret, holy, or sacred names to the uninitiated, or to obscure the identity of the subject of a conversation from potential eavesdroppers. Some Euphemisms are intended to amuse.

Synonyms

Synonyms are different words with identical or very similar meanings. Words that are Synonyms are said to be synonymous, and the state of being a synonym is called synonymy. The word comes from Ancient Greek syn (`with`) and onoma (á½„νομα) (`name`).

Tone

Tone is a literary technique that is a part of composition, which encompasses the attitudes toward the subject and toward the audience implied in a literary work. Tone may be formal, informal, intimate, solemn, somber, playful, serious, ironic, condescending, or many other possible attitudes. Tone and mood are not interchangeable.

Personal pronouns	Personal pronouns are pronouns used as substitutes for proper or common nouns. All known human languages have Personal pronouns. English in common use today has seven Personal pronouns: · first-person singular · first-person plural (we) · second-person singular and plural · third-person singular human or animate male (he) · third-person singular human or animate female (she) · third-person singular inanimate (it) · third-person plural Each pronoun has up to five forms: · A form used as the subject of a finite verb (I/we) as cited above. · An objective or oblique form (me/us/etc). used as the object of a verb or of a preposition · A reflexive form (myself/ourselves/etc). which replaces the objective-case form in referring to the same entity as the subject. · And two possessive forms (my/our/etc.

Chapter 2. PART II: Chapter 9 - Chapter 16

Language	Language may refer either to the specifically human capacity for acquiring and using complex systems of communication, or to a specific instance of such a system of complex communication. The scientific study of language in any of its senses is called linguistics. The approximately 3000-6000 languages that are spoken by humans today are the most salient examples, but natural languages can also be based on visual rather than auditive stimuli, for example in sign languages and written language.
Colloquialism	A Colloquialism is an expression not used in formal speech, writing or paralinguistics. Colloquialisms are also sometimes referred to collectively as 'colloquial language'. Colloquialisms or colloquial language is considered to be characteristic of or only appropriate for casual, ordinary, familiar, or informal conversation rather than formal speech or writing.
Slang	Slang is the use of highly informal words and expressions that are not considered standard in the speaker`s dialect or language. It is often used as a way to say words that are not appropriate, and it is also used to identify with others that are similar to you. Few linguists have endeavored to clearly define what constitutes Slang.
Communication	Communication is a process of transferring information from one entity to another. Communication processes are sign-mediated interactions between at least two agents which share a repertoire of signs and semiotic rules. Communication is commonly defined as `the imparting or interchange of thoughts, opinions, or information by speech, writing, or signs`.
Grammar	In linguistics, Grammar is the set of structural rules that govern the composition of sentences, phrases, and words in any given natural language. The term refers also to the study of such rules, and this field includes morphology, syntax, and phonology, often complemented by phonetics, semantics, and pragmatics. Linguists do not normally use the term to refer to orthographical rules, although usage books and style guides that call themselves Grammars may also refer to spelling and punctuation.
Revision	Revision is the stage in the writing process where the author reviews, alters, and amends her or his message, according to what has been written in the draft. Revision follows drafting and precedes editing. Drafting and revising often form a loop as a work moves back and forth between the two stages.
Chart	A Chart is a visual representation of data, in which the data are represented by symbols such as bars in a bar Chart or lines in a line Chart. A Chart can represent tabular numeric data, functions or some kinds of qualitative structures.

The term `Chart` as a visual representation of data has multiple meanings.

Flowchart

A Flowchart is a common type of diagram, that represents an algorithm or process, showing the steps as boxes of various kinds, and their order by connecting these with arrows. Flowcharts are used in analyzing, designing, documenting or managing a process or program in various fields. The first structured method for documenting process flow, the 'flow process chart', was introduced by Frank Gilbreth to members of ASME in 1921 as the presentation 'Process Charts--First Steps in Finding the One Best Way'.

Pictogram

A pictograph (also called pictogram is an ideogram that conveys its meaning through its pictorial resemblance to a physical object. Earliest examples of pictographs include ancient or prehistoric drawings or paintings found on rock walls. Pictographs are also used in writing and graphic systems in which the characters are to considerable extent pictorial in appearance.

Prose

Prose is the most typical form of written language, applying ordinary grammatical structure and natural flow of speech rather than rhythmic structure (as in traditional poetry). The English word "prose" is derived from the Latin prosa, which literally translates as "straight-forward." While there are critical debates on the construction of prose, its simplicity and loosely defined structure has led to its adoption for the majority of spoken dialogue, factual discourse as well as topical and fictional writing. It is commonly used, for example, in literature, newspapers, magazines, encyclopedias, broadcasting, film, history, philosophy, law and many other forms of communication.

Graphics

Graphics are visual presentations on some surface, such as a wall, canvas, computer screen, paper, inform, illustrate, or entertain. Examples are photographs, drawings, Line Art, graphs, diagrams, typography, numbers, symbols, geometric designs, maps, engineering drawings, or other images. Graphics often combine text, illustration, and color.

Photograph

A Photograph is an image created by light falling on a light-sensitive surface, usually Photographic film or an electronic imager such as a CCD or a CMOS chip. Most Photographs are created using a camera, which uses a lens to focus the scene`s visible wavelengths of light into a reproduction of what the human eye would see. The process and practice of creating Photographs is called Photography.

Schematic	A Schematic is a diagram that represents the elements of a system using abstract, graphic symbols rather than realistic pictures. A Schematic usually omits all details that are not relevant to the information the Schematic is intended to convey, and may add unrealistic elements that aid comprehension. For example, a subway map intended for riders may represent a subway station with a dot; the dot doesn`t resemble the actual station at all but gives the viewer information without unnecessary visual clutter.
Line	A line is a unit of language into which a poem or play is divided, which operates on principles which are distinct from and not necessarily coincident with grammatical structures, such as the sentence or clauses in sentences. Although the word for a single poetic line is verse, that term now tends to be used to signify poetic form more generally. A distinct numbered group of lines in verse is normally called a stanza.
Thumbnail	Thumbnails are reduced-size versions of pictures, used to help in recognizing and organizing them, serving the same role for images as a normal text index does for words. In the age of digital images, visual search engines and image-organizing programs normally use Thumbnails, as do most modern operating systems or desktop environments, such as Microsoft Windows, Mac OS X, KDE, and GNOME. Some web designers produce Thumbnails with HTML coding that makes the user`s browser shrink the picture, rather than use a smaller copy of the image. In principle the display size of an image in pixels should always correspond to its actual size, in part because one purpose of a Thumbnail image on a web page is to reduce download time.
Brief	An architectural brief is, in its broadest sense, a requirement a client may have that an architect designs to meet, usually by creating a building to accommodate the requirement(s). A brief is a written document that might be anything from a single page to a multiple volume set of documents. It is often formulated by or with the architect, and includes opportunities assessed, constraints identified and needs for further investigation clarified.
Image	An Image is an artifact, that has a similar appearance to some subject--usually a physical object or a person. Images may be two-dimensional, such as a photograph, screen display, and as well as a three-dimensional, such as a statue. They may be captured by optical devices--such as cameras, mirrors, lenses, telescopes, microscopes, etc.

Desktop publishing	Desktop publishing (also known as DTP) combines a personal computer and WYSIWYG page layout software to create publication documents on a computer for either large scale publishing or small scale local multifunction peripheral output and distribution. The term 'Desktop publishing' is commonly used to describe page layout skills. However, the skills and software are not limited to paper and book publishing.
Guide	A Guide is a person who leads anyone through unknown or unmapped country. This includes a Guide of the real world (such as someone who conducts travellers and tourists through a place of interest), as well as a person who leads someone to more abstract places (such as to knowledge or wisdom). The word 'Guide' was incorporated into English via Old French 'Guider' which meant 'to Guide, lead, conduct' it was originally taken by Old French from Frankish '*witan' meaning 'show the way' from Proto-Germanic '*wit-' meaning 'to know' .
Language	Language may refer either to the specifically human capacity for acquiring and using complex systems of communication, or to a specific instance of such a system of complex communication. The scientific study of language in any of its senses is called linguistics. The approximately 3000-6000 languages that are spoken by humans today are the most salient examples, but natural languages can also be based on visual rather than auditive stimuli, for example in sign languages and written language.
Skills	A skill is the learned capacity to carry out pre-determined results often with the minimum outlay of time, energy, or both. skills can often be divided into domain-general and domain-specific skills. For example, in the domain of work, some general skills would include time management, teamwork and leadership, self motivation and others, whereas domain-specific skills would be useful only for a certain job.
Style guide	A Style guide or style manual is a set of standards for design and writing of documents, either for general use or for a specific publication or organization. Style guides are prevalent for general and specialized use, for the general reading and writing audience, and for students and scholars of the various academic disciplines, medicine, journalism, the law, government, business, and industry. Some Style guides focus on graphic design, covering such topics as typography and white space.

Paper

The Chinese invention of paper and the advent of woodblock printing produced the world`s first print culture. As the print scholar A. Hyatt Mayor noted, `it was the Chinese who really discovered the means of communication that was to dominate until our age.` Another scholar, the prominent American historian and critic Daniel J. Boorstin, also noted in his book The Discoverers, singled out Korea as the `most developed nation in the field of printing.`
Woodblock printing was better suited to Chinese characters than movable type, which the Chinese also invented, but which did not replace woodblock printing. In China and Korea, the use of woodblock printing on paper and movable type preceded their use in Europe by several centuries.

Space

In writing, a space is a blank area devoid of content, serving to separate words, letters, numbers, and punctuation. Conventions for interword and intersentence spaces vary among languages, and in some cases the spacing rules are quite complex.

In the classical period, Latin was written with interpuncts (centred dots) as word separators, but that practice was abandoned sometime around the year 200 CE in favour of scriptio continua, i.e., with the words running together without any word separators.

Rights

Rights are normative principles, variously construed as legal, social, or moral freedoms or entitlements. While the concept of Rights is fundamental to civilized societies, there is considerable disagreement about what is meant precisely by the term Rights. It has been used by different groups and thinkers for different purposes, with different and sometimes opposing definitions, and the precise definition of the concept, beyond having something to do with normative rules of some sort or another, is controversial.

Serif

In typography, Serifs are semi-structural details on the ends of some of the strokes that make up letters and symbols. A typeface that has Serifs is called a Serif typeface (or Seriffed typeface). A typeface without Serifs is called sans-Serif, from the French sans, meaning `without`.

Typeface

In typography, a typeface is a set of one or more fonts, in one or more sizes, designed with stylistic unity, each comprising a coordinated set of glyphs. A typeface usually comprises an alphabet of letters, numerals, and punctuation marks; it may also include ideograms and symbols, or consist entirely of them, for example, mathematical or map-making symbols. The term typeface is frequently conflated with font; the two terms had more clearly differentiated meanings before the advent of desktop publishing.

Typography

Typography is the art and technique of arranging type, type design, and modifying type glyphs. Type glyphs are created and modified using a variety of illustration techniques. The arrangement of type involves the selection of typefaces, point size, line length, leading (line spacing), adjusting the spaces between groups of letters (tracking) and adjusting the space between pairs of letters (kerning).

Phrase

In grammar, a Phrase is a group of words functioning as a single unit in the syntax of a sentence.

For example, the store at the end of the street is a Phrase. It acts like a noun.

Business letter

A commercial business letter is a letter written in formal language, usually used when writing from one business organization to another, or for correspondence between such organizations and their customers, clients and other external parties. The overall style of letter will depend on the relationship between the parties concerned.

General Format

There are two main styles of business letters:

1. Full block style: Align all elements on the left margin.
2. Modified block style: Down the middle of the page, align the return address, date, closing, signature, and typed name; align other elements on the left page margin.

Elements

Business letters (in the United States) usually contain the following elements, in order:

- Sender's address ' contact information
- Date of writing
- Recipient's name, title, company, ' address
- Salutation/greeting
- Subject
- Message (body of the letter)
- Valediction/closing
- Sender's signature
- Sender's name, title, company

In some situations, a business letter may also include the following optional information:

- Enclosures (Encl.: or Enc.:)
- Carbon Copy Recipients (cc:)
- Reference Initials (of the typist, if different from original author of letter)

Line Spacing

In general, each element or paragraph of the letter is followed by a single blank line, except:

- the date, followed by three or four blank lines;
- the final content paragraph, followed by two blank lines;
- the valediction/closing, followed by three or four blank lines (enough for the sender to sign the letter); and
- the sender's title, followed by two blank lines.

Font Formatting

No special character or font formatting is used, except for the subject line, which is usually underlined

Punctuation

	The salutation/greeting is generally followed by a comma, although in the United States a colon is often preferred.
Digital	A digital system is a data technology that uses discrete (discontinuous) values. By contrast, non-digital (or analog) systems use a continuous range of values to represent information. Although digital representations are discrete, the information represented can be either discrete, such as numbers, letters or icons, or continuous, such as sounds, images, and other measurements of continuous systems.
Recommendation	A Recommendation can be: · In technical contexts, a norm (philosophy) or a norm (sociology) · European Union Recommendation, in international law · W3C Recommendation, in Internet contexts · a letter of Recommendation in employment or academia
Meeting	In a Meeting, two or more people come together for the purpose of discussing a (usually) predetermined topic such as business or community event planning, often in a formal setting. In addition to coming together physically (in real life, face to face), communication lines and equipment can also be set up to have a discussion between people at different locations, e.g. a conference call or an e-Meeting. In organizations, Meetings are an important vehicle for personal contact.
Minutes	Minutes are the instant written record of a meeting or hearing. They often give an overview of the structure of the meeting, starting with a list of those present, a statement of the various issues before the participants, and each of their responses thereto. They are often created at the moment of the hearing by a typist or court recorder at the meeting, who may record the meeting in shorthand, and then prepare the Minutes and issue them to the participants afterwards.

Copyright	Copyright is the set of exclusive rights granted to the author or creator of an original work, including the right to copy, distribute and adapt the work. These rights can be licensed, transferred and/or assigned. Copyright lasts for a certain time period after which the work is said to enter the public domain.
Electronic privacy	The protection of electronic mail from unauthorized access and inspection is known as electronic privacy. In countries with a constitutional guarantee of the secrecy of correspondence, e-mail is equated with letters and thus legally protected from all forms of eavesdropping. In the United States, privacy of correspondence is derived from the Fourth Amendment to the United States Constitution and thus restricted by the requirement for a `reasonable expectation of privacy`.

Alternative media are media (newspapers, radio, television, movies, Internet, etc). which are alternatives to the business or government-owned mass media. Proponents of Alternative media argue that the mainstream media are biased.

Alternative media	Newspapers · Magazines · News agencies · Broadcast · Online · Photojournalism · Alternative media

Roles

Journalist · Marketer · Reporter · Editor · Columnist · Commentator · Photographer · Presenter · Meteorologist · Production Manager · Intern

This box: view · talk · edit

Instant messaging	Instant messaging is a form of real-time communication between two or more people based on typed text. The text is conveyed via devices connected over a network such as the Internet. Instant messaging is a collection of technologies that create the possibility of real-time text-based communication between two or more participants over the internet or some form of internal network/intranet.

Collaboration	Collaboration is a recursive process where two or more people or organizations work together in an intersection of common goals -- for example, an intellectual endeavor that is creative in nature--by sharing knowledge, learning and building consensus. Most Collaboration requires leadership, although the form of leadership can be social leadership within a decentralized and egalitarian group. In particular, teams that work collaboratively can obtain greater resources, recognition and reward when facing competition for finite resources.Collaboration is also present in opposing goals exhibiting the notion of adversarial Collaboration, though this is not a common case for using the term.

Letter

A letter is a written message from one party to another. The role of letters in communication has changed significantly since the 19th century. Historically, letters (in paper form) were the only reliable means of communication between two persons in different locations.

Adjustment

Adjustment means regulating, adapting or settling in a variety of contexts:
Adjustment has several meanings; many relate to insurance, contracts, or the resolution of disputes.

In Engineering, Mathematics and Geodesy, Adjustment means the optimal parameter estimation of a mathematical model so as to best fit a data set. The most important method is the least squares Adjustment, found by Carl Friedrich Gauss.

Address

An address is a collection of information, presented in a mostly fixed format, used for describing the location of a building, apartment, or other structure or a plot of land, generally using political boundaries and street names as references, along with other identifiers such as house or apartment numbers. Some addresses also contain special codes to aid routing of mail and packages, such as a ZIP code or post code.

Functions

Addresses have several functions:

1. Providing a means of physically locating a building, especially in a city where there are many buildings and streets,
2. Identifying buildings as the end points of a postal system,
3. A social function: someone's address can have a profound effect on their social standing,
4. As parameters in statistics collection, especially in census-taking or the insurance industry.

History

Until the advent of modern postal systems, most houses and buildings were not numbered.

Salutation

A salutation is a greeting, in particular a formal greeting used in a letter. Salutations usually take the form "Dear [recipient's given name]". For each style of salutation there is an accompanying style of complimentary close (valediction).

Business letter

A commercial business letter is a letter written in formal language, usually used when writing from one business organization to another, or for correspondence between such organizations and their customers, clients and other external parties. The overall style of letter will depend on the relationship between the parties concerned.

General Format

There are two main styles of business letters:

1. Full block style: Align all elements on the left margin.
2. Modified block style: Down the middle of the page, align the return address, date, closing, signature, and typed name; align other elements on the left page margin.

Elements

Business letters (in the United States) usually contain the following elements, in order:

- Sender's address ' contact information
- Date of writing
- Recipient's name, title, company, ' address
- Salutation/greeting
- Subject
- Message (body of the letter)
- Valediction/closing
- Sender's signature
- Sender's name, title, company

In some situations, a business letter may also include the following optional information:

- Enclosures (Encl.: or Enc.:)
- Carbon Copy Recipients (cc:)
- Reference Initials (of the typist, if different from original author of letter)

Line Spacing

In general, each element or paragraph of the letter is followed by a single blank line, except:

- the date, followed by three or four blank lines;
- the final content paragraph, followed by two blank lines;
- the valediction/closing, followed by three or four blank lines (enough for the sender to sign the letter); and
- the sender's title, followed by two blank lines.

Font Formatting

No special character or font formatting is used, except for the subject line, which is usually underlined

Punctuation

The salutation/greeting is generally followed by a comma, although in the United States a colon is often preferred.

Signature

A signature is a handwritten (and sometimes stylized) depiction of someone's name, nickname, or even a simple "X" that a person writes on documents as a proof of identity and intent. The writer of a signature is a signatory. Similar to a handwritten signature, a signature work describes the work as readily identifying its creator.

Signature block

A Signature block is a block of text automatically appended at the bottom of an e-mail message, Usenet article, or forum post. This has the effect of `signing off` the message and in a reply message of indicating that no more response follows. It is common practice for a Signature block to consist of one or more lines containing some brief information on the author of the message.

Attention

Attention is the cognitive process of selectively concentrating on one aspect of the environment while ignoring other things. Attention has also been referred to as the allocation of processing resources

Examples include listening carefully to what someone is saying while ignoring other conversations in a room (the cocktail party effect) or listening to a cell phone conversation while driving a car. Attention is one of the most intensely studied topics within psychology and cognitive neuroscience.

Notations

This page is an overview of the common notations used when working with vectors, which may be spatial or more abstract members of vector spaces.

The common typographic convention for representing a vector is upright boldface type, as in v for a vector named `v`. In handwriting, where boldface type is either unavailable or unwieldy, vectors are often represented with right-pointing arrows or harpoons above their names, as in  .

PostScript

A postscript, abbreviated P.S., is writing added after the main body of a letter (or other body of writing). The term comes from the Latin post scriptum, an expression meaning "written after" (which may be interpreted in the sense of "that which comes after the writing").

A postscript may be a sentence, a paragraph, or occasionally many paragraphs added, often hastily and incidentally, after the signature of a letter or (sometimes) the main body of an essay or book.

Stationery

Stationery has historically meant a wide gamut of materials: paper and office supplies, writing implements, greeting cards, glue, pencil case etc.

Originally the term `Stationery` referred to all products sold by a stationer, whose name indicates that his book shop was on a fixed spot, usually near a university, and permanent, while medieval trading was mainly ambulant, by peddlers (including chapmen, who sold books) and others (such as farmers and craftsmen) at non-permanent markets such as fairs. It was a special term used between the 13th and 15th centuries in the manuscript culture.

Plain English

Plain English is a generic term for communication styles that emphasise clarity, brevity and the avoidance of technical language.

Plain English is written in a manner appropriate to the reading skills and knowledge of its audience. It is writing that has no confusion about meaning, is free of cliché and unnecessary jargon, and is easy to understand.

Business letter

A commercial business letter is a letter written in formal language, usually used when writing from one business organization to another, or for correspondence between such organizations and their customers, clients and other external parties. The overall style of letter will depend on the relationship between the parties concerned.

General Format

There are two main styles of business letters:

1. Full block style: Align all elements on the left margin.
2. Modified block style: Down the middle of the page, align the return address, date, closing, signature, and typed name; align other elements on the left page margin.

Elements

Business letters (in the United States) usually contain the following elements, in order:

- Sender's address ' contact information
- Date of writing
- Recipient's name, title, company, ' address
- Salutation/greeting
- Subject
- Message (body of the letter)
- Valediction/closing
- Sender's signature
- Sender's name, title, company

In some situations, a business letter may also include the following optional information:

- Enclosures (Encl.: or Enc.:)
- Carbon Copy Recipients (cc:)
- Reference Initials (of the typist, if different from original author of letter)

Line Spacing

In general, each element or paragraph of the letter is followed by a single blank line, except:

- the date, followed by three or four blank lines;
- the final content paragraph, followed by two blank lines;
- the valediction/closing, followed by three or four blank lines (enough for the sender to sign the letter); and
- the sender's title, followed by two blank lines.

Font Formatting

No special character or font formatting is used, except for the subject line, which is usually underlined

Punctuation

The salutation/greeting is generally followed by a comma, although in the United States a colon is often preferred.

Complaint

In legal terminology, a complaint is a formal legal document that sets out the facts and legal reasons that the filing party or parties (the plaintiff(s)) believes are sufficient to support a claim against the party or parties against whom the claim is brought (the defendant(s)) that entitles the plaintiff(s) to a remedy (either money damages or injunctive relief). For example, the Federal Rules of Civil Procedure that govern civil litigation in United States courts provide that a civil action is commenced with the filing or service of a pleading called a complaint. Civil court rules in states that have incorporated the Federal Rules of Civil Procedure use the same term for the same pleading.

Telephone	The Telephone is a telecommunications device that transmits and receives sound, most commonly the human voice. It is one of the most common household appliances in the developed world, and has long been considered indispensable to business, industry and government. The word `Telephone` has been adapted to many languages and is widely recognized around the world.
Skills	A skill is the learned capacity to carry out pre-determined results often with the minimum outlay of time, energy, or both. skills can often be divided into domain-general and domain-specific skills. For example, in the domain of work, some general skills would include time management, teamwork and leadership, self motivation and others, whereas domain-specific skills would be useful only for a certain job.
Blog	A blog is a type of website or part of a website. Blogs are usually maintained by an individual with regular entries of commentary, descriptions of events, or other material such as graphics or video. Entries are commonly displayed in reverse-chronological order.
Organizations	Organizations · ALLARM · National Coalition on Racism in Sports and Media · STAR · Honor Indians Institute Articles · `Crimes Against Humanity` by Ward Churchill Parody · Turning the Tables on Racist Mascots

· Alternative Indian Mascots

Recommendation

A Recommendation can be:

· In technical contexts, a norm (philosophy) or a norm (sociology)

· European Union Recommendation, in international law

· W3C Recommendation, in Internet contexts

· a letter of Recommendation in employment or academia

Recommendation letter

A letter of recommendation is a letter in which the writer assesses the qualities, characteristics, and capabilities of the person being recommended in terms of that individual's ability to perform a particular task or function. Recommendation letters are almost always specifically requested to be written about someone, and are therefore addressed to a particular requestor. Letters of recommendation are typically related to employment, admissions to institutions of higher education or scholarship eligibility.

Organizations

Organizations

· ALLARM

· National Coalition on Racism in Sports and Media

· STAR

· Honor Indians Institute

Articles

· `Crimes Against Humanity` by Ward Churchill

	Parody · Turning the Tables on Racist Mascots · Alternative Indian Mascots
Internet	The Internet is a global system of interconnected computer networks that use the standardized Internet Protocol Suite (TCP/IP) to serve billions of users worldwide. It is a network of networks that consists of millions of private and public, academic, business, and government networks of local to global scope that are linked by copper wires, fiber-optic cables, wireless connections, and other technologies. The Internet carries a vast array of information resources and services, most notably the inter-linked hypertext documents of the World Wide Web (WWW) and the infrastructure to support electronic mail.
Source	A source or sender is one of the basic concepts of communication and information processing. sources are objects which encode message data and transmit the information, via a channel, to one or more observers (or receivers). In the strictest sense of the word, particularly in information theory, a source is a process that generates message data that one would like to communicate, or reproduce as exactly as possible somewhere else in space or time.
Web browser	A Web browser is a software application for retrieving, presenting, and traversing information resources on the World Wide Web. An information resource is identified by a Uniform Resource Identifier (URI) and may be a web page, image, video, or other piece of content. Hyperlinks present in resources enable users to easily navigate their browsers to related resources.
Usability	Usability is the ease of use and learnability of a human-made object. The object of use can be a software application, website, book, tool, machine, process, or anything a human interacts with. A usability study may be conducted as a primary job function by a usability analyst or as a secondary job function by designers, technical writers, marketing personnel, and others.
Audience	An Audience is a group of people who participate in a show or encounter a work of art, literature (in which they are called the 'reader'), theatre, music or academics in any medium. Audience members participate in different ways in different kinds of art; some events invite overt Audience participation and others allowing only modest clapping and criticism and reception. Media Audiences are studied by academics in media Audience studies.

Employee

Employment is a contract between two parties, one being the employer and the other being the employee. An employee may be defined as: `A person in the service of another under any contract of hire, express or implied, oral or written, where the employer has the power or right to control and direct the employee in the material details of how the work is to be performed.` Black`s Law Dictionary page 471 (5th ed. 1979).

In a commercial setting, the employer conceives of a productive activity, generally with the intention of generating a profit, and the employee contributes labour to the enterprise, usually in return for payment of wages.

Employee

Employment is a contract between two parties, one being the employer and the other being the employee. An employee may be defined as: `A person in the service of another under any contract of hire, express or implied, oral or written, where the employer has the power or right to control and direct the employee in the material details of how the work is to be performed.` Black`s Law Dictionary page 471 (5th ed. 1979).

In a commercial setting, the employer conceives of a productive activity, generally with the intention of generating a profit, and the employee contributes labour to the enterprise, usually in return for payment of wages.

Communication

Communication is a process of transferring information from one entity to another. Communication processes are sign-mediated interactions between at least two agents which share a repertoire of signs and semiotic rules. Communication is commonly defined as `the imparting or interchange of thoughts, opinions, or information by speech, writing, or signs`.

Technical definition

A Technical definition is a definition in technical communication describing or explaining technical terminology. Technical definitions are used to introduce the vocabulary which makes communication in a particular field succinct and unambiguous. (For example, the illiac crest from medical terminology is the top ridge of the hip bone.

Sentence

In the field of linguistics, a sentence is an expression in natural language, and often defined to indicate a grammatical unit consisting of one or more words that generally bear minimal syntactic relation to the words that precede or follow it. A sentence can include words grouped meaningfully to express a statement, question, exclamation, request, command or suggestion.

	As with all language expressions, sentences may contain both function and content words, and contain properties distinct to natural language, such as characteristic intonation and timing patterns.
Acronyms	Acronyms and initialisms are abbreviations that are formed using the initial components in a phrase or name. These components may be individual letters (as in CEO) or parts of words (as in Benelux). There is no universal agreement on the precise definition of the various terms , nor on written usage .
Etymology	Etymology is the study of the history of words and how their form and meaning have changed over time. Etymology is that branch of linguistics, which deals with the study and application of history of words, their formation and meaning that have changed over time. For languages with a long written history, etymologists make use of texts in these languages, and texts about the languages, to gather knowledge about how words were used at earlier stages, and when they entered the languages in question.
Jargon	Jargon is terminology which is especially defined in relationship to a specific activity, profession, group, or event. The philosophe Condillac observed in 1782 that "Every science requires a special language because every science has its own ideas." As a rationalist member of the Enlightenment he continued, "It seems that one ought to begin by composing this language, but people begin by speaking and writing and the language remains to be composed." In other words, the term covers the language used by people who work in a particular area or who have a common interest. Much like slang, it can develop as a kind of short-hand, to express ideas that are frequently discussed between members of a group, though it can also be developed deliberately using chosen terms.
Contrast	In literature, an author writes contrast when he or she describes the difference(s) between two or more entities. For example, in the first four lines of William Shakespeare's Sonnet 130, Shakespeare contrasts a mistress to the sun, coral, snow, and wire.

Contrast is the antonym of simile.

Editing

Editing is the process of preparing language, images, sound, video condensation, organization, and other modifications in various media. A person who edits is called an editor. In a sense, the Editing process originates with the idea for the work itself and continues in the relationship between the author and the editor.

Writing

Writing is the representation of language in a textual medium through the use of a set of signs or symbols (known as a writing system). It is distinguished from illustration, such as cave drawing and painting, and non-symbolic preservation of language via non-textual media, such as magnetic tape audio.

Writing most likely began as a consequence of political expansion in ancient cultures, which needed reliable means for transmitting information, maintaining financial accounts, keeping historical records, and similar activities.

Writing style

Writing style is the manner in which a writer chooses among different strategies to address an issue and an audience. A style reveals the writer's personality or voice, but it also shows how she or he sees the audience of the writing. The writing style reveals the choices the writer makes in syntactical structures, diction, and figures of thought.

Case study

A case study is one of several ways of doing research whether it is social science related or even socially related. It is an in-depth investigation/study of a single individual, group, incident, or community. Other ways include experiments, surveys, or analysis of archival information .

Description

Description is one of four rhetorical modes (also known as modes of discourse), along with exposition, argumentation, and narration. Each of the rhetorical modes is present in a variety of forms and each has its own purpose and conventions.
Description is also the fiction-writing mode for transmitting a mental image of the particulars of a story.

Description

Description is one of four rhetorical modes (also known as modes of discourse), along with exposition, argumentation, and narration. Each of the rhetorical modes is present in a variety of forms and each has its own purpose and conventions.

Description is also the fiction-writing mode for transmitting a mental image of the particulars of a story.

Outline

An outline is a list of the main features of a given topic, often used as a rough draft or summary of the content of a document. A hierarchical outline is a list arranged to show hierarchical relationships, and is a type of tree structure.

Writers of fiction and creative nonfiction, such as Jon Franklin, may use outlines to establish plot sequence, character development and dramatic flow of a story, sometimes in conjunction with freewriting.

Specification

A specification is an explicit set of requirements to be satisfied by a material, product, or service. Should a material, product or service fail to meet one or more of the applicable specifications, it may be referred to as being out of specification; the abbreviation OOS may also be used.

A technical specification may be developed privately, for example by a corporation, regulatory body, or military organization, or it may be developed by standards organizations which often have more diverse input and usually develop voluntary standards (Voluntary standards may become mandatory if adopted by a government or business contract).

Specification

A specification is an explicit set of requirements to be satisfied by a material, product, or service. Should a material, product or service fail to meet one or more of the applicable specifications, it may be referred to as being out of specification; the abbreviation OOS may also be used.

A technical specification may be developed privately, for example by a corporation, regulatory body, or military organization, or it may be developed by standards organizations which often have more diverse input and usually develop voluntary standards (Voluntary standards may become mandatory if adopted by a government or business contract).

Literature	Literature is the art of written works, and is not bound to published sources (although, under circumstances they can be exempt). Literally translated, the word literature means "acquaintance with letters" (as in the "arts and letters"). The two most basic written literary categories include fiction and non fiction.
Guide	A Guide is a person who leads anyone through unknown or unmapped country. This includes a Guide of the real world (such as someone who conducts travellers and tourists through a place of interest), as well as a person who leads someone to more abstract places (such as to knowledge or wisdom). The word 'Guide' was incorporated into English via Old French 'Guider' which meant 'to Guide, lead, conduct' it was originally taken by Old French from Frankish '*witan' meaning 'show the way' from Proto-Germanic '*wit-' meaning 'to know' .
Fact sheet	A Fact sheet is a presentation of data on any subject in a format emphasizing brevity, key points of interest or concern, a fairly minimalist design aesthetic, and a general desire to convey the most relevant information in the least amount of space. As such, it is a medium ideally suited to the post-modern information age, when `facts` (to the extent they exist) proliferate in ever-increasing numbers while the time available to review and comprehend them diminishes. Fact sheets attempt to solve this problem by condensing information into an easy-to-read, straightforward, portable, and modular form of knowledge.
Technical communication	Technical communication is the process of conveying technical information through writing, speech, and other mediums to a specific audience. Information is usable if the intended audience can perform an action or make a decision based on it (Johnson-Sheehan 7). Technical communicators often work collaboratively to create products (deliverables) for various media, including paper, video, and the Internet.
Copyright	Copyright is the set of exclusive rights granted to the author or creator of an original work, including the right to copy, distribute and adapt the work. These rights can be licensed, transferred and/or assigned. Copyright lasts for a certain time period after which the work is said to enter the public domain.
Reference card	A Reference card or quick Reference card is a concise bundling of common syntactic rules and idioms of a particular formal language. It serves as an ad hoc memory aid for an experienced user. In spite of what the name Reference card suggest, the term only tends to be used in the narrow context of programming languages or markup languages.

Information	Information as a concept has many meanings, from everyday usage to technical settings. The concept of Information is closely related to notions of constraint, communication, control, data, form, instruction, knowledge, meaning, mental stimulus, pattern, perception, and representation. In its most restricted technical meaning, Information is an ordered sequence of symbols.
Notes	Notes · The negation particles are na and nahi with the former standing before the copula (or if no copula, the aspectual form) and the latter generally after. A negation particle combines with present ch-PN however for the invariable nathÄ«. An alternative to the past na hat-GN is nahot-GN. · Gujarati retains an aspectually unmarked form in the function of the Present Imperfective, although a marked form (hÄ lto nathÄ«) replaces it in the negative.
Readability	Readability can mean how easy it is to read and to make sense of a text. Or it can mean how easy it is to recognise words on a page, sign, or display. In English, this is often called legibility.
Esophageal speech	Esophageal speech (or voice) is an alternate method for speech production without the oscillation in the vocal folds. The sound is produced by releasing gases from or through the esophagus. In Esophageal speech it is thus the esophagus that oscillates in contrast to normal (laryngeal) speech where the vocal folds oscillate.
Verb	In syntax, a Verb, deriving from the Latin Verbum meaning word, is a word that conveys action (bring, read, walk, run, murder), or a state of being (exist, stand). In most languages, Verbs are inflected (modified in form) to encode tense, aspect, mood and voice. A Verb may also agree with the person, gender, and/or number of some of its arguments, such as its subject, or object.
Introduction	An introduction is a beginning section which states the purpose and goals of the following writing. The introduction is usually interesting and it intrigues the reader and causes him or her to want to read on. The sentence in which the introduction begins can be a question or just a statement.

Documentation

Documentation is a general term for a multiplicity of documents in a chosen mix of media and with a certain collection. Purpose of documentation is the use to support a tool or a process.

Classical documentation is a set of documents printed on paper.

Focus group

A Focus group is a form of qualitative research in which a group of people are asked about their attitude towards a product, service, concept, advertisement, idea, or packaging. Questions are asked in an interactive group setting where participants are free to talk with other group members.

The first Focus groups were created at the Bureau of Applied Social Research by associate director, sociologist Robert K. Merton.

Double bind

A Double bind is a dilemma in communication in which an individual receives two or more conflicting messages, with one message negating the other. This creates a situation in which a successful response to one message implicates a failed response to the other, so that the person will be automatically wrong regardless of response. The person can neither comment on the conflict, nor resolve it, nor opt out of the situation.

Double bind

A Double bind is a dilemma in communication in which an individual receives two or more conflicting messages, with one message negating the other. This creates a situation in which a successful response to one message implicates a failed response to the other, so that the person will be automatically wrong regardless of response. The person can neither comment on the conflict, nor resolve it, nor opt out of the situation.

Classroom

A classroom is a room in which teaching or learning activities can take place. classrooms are found in educational institutions of all kinds, including public and private schools, corporations, and religious and humanitarian organizations. The classroom attempts to provide a safe space where learning can take place uninterrupted by other distractions.

Minutes

Minutes are the instant written record of a meeting or hearing. They often give an overview of the structure of the meeting, starting with a list of those present, a statement of the various issues before the participants, and each of their responses thereto. They are often created at the moment of the hearing by a typist or court recorder at the meeting, who may record the meeting in shorthand, and then prepare the Minutes and issue them to the participants afterwards.

Procedural

A Procedural is a cross-genre type of literature, film, or television program involving a sequence of technical detail. A documentary film may be written in a procedural style to heighten narrative interest.

Types

Television

Fiction

In television, "procedural" specifically refers to a genre of programs in which a problem is introduced, investigated and solved all within the same episode.

Proposals

Proposals is a play by Neil Simon. Incidental music was written by Stephen Flaherty.

A nostalgic memory play, proposals recalls one idyllic afternoon in the summer of 1953, the last time the Hines clan gathers at its retreat in the Poconos.

Proposals

Proposals is a play by Neil Simon. Incidental music was written by Stephen Flaherty.

A nostalgic memory play, proposals recalls one idyllic afternoon in the summer of 1953, the last time the Hines clan gathers at its retreat in the Poconos.

Documentation

Documentation is a general term for a multiplicity of documents in a chosen mix of media and with a certain collection. Purpose of documentation is the use to support a tool or a process.

Classical documentation is a set of documents printed on paper.

Supplement

A supplement is a publication that has a role secondary to that of another preceding or concurrent publication.

A follow-on publication complements its predecessor, either by bringing it up-to-date (e.g. the Index Catalogue), or by otherwise enhancing the predecessor's coverage of a particular topic or subject matter, as in the Tosefta. Supplements are particularly popular and useful in gaming hobbies.

Abstract

An abstract is a brief summary of a research article, thesis, review, conference proceeding or any in-depth analysis of a particular subject or discipline, and is often used to help the reader quickly ascertain the paper's purpose. When used, an abstract always appears at the beginning of a manuscript, acting as the point-of-entry for any given scientific paper or patent application. Abstracting and indexing services for various academic disciplines are aimed at compiling a body of literature for that particular subject.

Comparative

In grammar, the Comparative is the form of an adjective , thing, or other entity has a property or quality greater or less in extent than that of another, and is used in this context with a subordinating conjunction, such as than, as...as, etc.

Statement

Statement may refer to:

- A kind of expression in language (linguistics)
- Statement declarative sentence that is either true or false
- Statement the smallest standalone element of an imperative programming language
- "Statement" (song), 2008 song by the Japanese band Boris
- Statements (album), 1962 album by jazz vibraphonist Milt Jackson
- Financial statements, formal records of the financial activities of a business, person, or other entity
- Mathematical statement, a statement in logic and mathematics
- Political statement, any act or nonverbal form of communication that is intended to influence a decision to be made for or by a group
- Press statement, written or recorded communication directed at members of the news media
- Statement of Special Educational Needs, outlining specific provision needed for a child in England
- a paper size also known as organizer L or half letter

.

Abstract	An abstract is a brief summary of a research article, thesis, review, conference proceeding or any in-depth analysis of a particular subject or discipline, and is often used to help the reader quickly ascertain the paper's purpose. When used, an abstract always appears at the beginning of a manuscript, acting as the point-of-entry for any given scientific paper or patent application. Abstracting and indexing services for various academic disciplines are aimed at compiling a body of literature for that particular subject.
Recommendation	A Recommendation can be: · In technical contexts, a norm (philosophy) or a norm (sociology) · European Union Recommendation, in international law · W3C Recommendation, in Internet contexts · a letter of Recommendation in employment or academia
Table of contents	A table of contents, usually headed simply "Contents" and abbreviated informally as Table of contents, is a list of the parts of a book or document organized in the order in which the parts appear. The contents usually includes the titles or descriptions of the first-level headers, such as chapter titles in longer works, and often includes second-level or section titles (A-heads) within the chapters as well, and occasionally even third-level titles (subsections or B-heads). The depth of detail in tables of contents depends on the length of the work, with longer works having less.
Executive summary	Executive summary is a term used in business for a short document that summarizes a longer report, proposal or group of related reports in such a way that readers can rapidly become acquainted with a large body of material without having to read it all. It will usually contain a brief statement of the problem or proposal covered in the major document(s), background information, concise analysis and main conclusions. It is intended as an aid to decision making by business managers.
Poster	A Poster is any piece of printed paper designed to be attached to a wall or vertical surface. Typically Posters include both textual and graphic elements, although a Poster may be either wholly graphical or wholly textual. Posters are designed to be both eye-catching and convey information.

Chart	A Chart is a visual representation of data, in which the data are represented by symbols such as bars in a bar Chart or lines in a line Chart. A Chart can represent tabular numeric data, functions or some kinds of qualitative structures. The term `Chart` as a visual representation of data has multiple meanings.
Eye contact	Eye contact is a meeting of the eyes between two individuals. In human beings, Eye contact is a form of nonverbal communication and is thought to have a large influence on social behavior. Coined in the early to mid-1960s, the term has come in the West to often define the act as a meaningful and important sign of confidence and social communication.
Plagiarism	Plagiarism, as defined in the 1995 Random House Compact Unabridged Dictionary, is the `use or close imitation of the language and thoughts of another author and the representation of them as one`s own original work.` Within academia, Plagiarism by students, professors, or researchers is considered academic dishonesty or academic fraud, and offenders are subject to academic censure, up to and including expulsion. In journalism, Plagiarism is considered a breach of journalistic ethics, and reporters caught plagiarizing typically face disciplinary measures ranging from suspension to termination of employment. Some individuals caught plagiarizing in academic or journalistic contexts claim that they plagiarized unintentionally, by failing to include quotations or give the appropriate citation.
Paraphrasing	Paraphrase is restatement of a text or passages, using other words. The term `paraphrase` derives via the Latin `paraphrasis` from the Greek para phraseïn, meaning `additional manner of expression`. The act of Paraphrasing is also called `paraphrasis.` A paraphrase typically explains or clarifies the text that is being paraphrased.
Common knowledge	Common knowledge is that what `everybody knows`, usually with reference to the community in which the term is used. The assertion that something is `Common knowledge` is sometimes associated with the fallacy argumentum ad populum . The fallacy essentially warns against assuming that just because everyone believes something is true does not make it so.

Knowledge	Knowledge is defined by the Oxford English Dictionary as expertise, and skills acquired by a person through experience or education; the theoretical or practical understanding of a subject, (ii) what is known in a particular field or in total; facts and information or (iii) awareness or familiarity gained by experience of a fact or situation. Philosophical debates in general start with Plato`s formulation of knowledge as `justified true belief`. There is however no single agreed definition of knowledge presently, nor any prospect of one, and there remain numerous competing theories.
Paper	The Chinese invention of paper and the advent of woodblock printing produced the world`s first print culture. As the print scholar A. Hyatt Mayor noted, `it was the Chinese who really discovered the means of communication that was to dominate until our age.` Another scholar, the prominent American historian and critic Daniel J. Boorstin, also noted in his book The Discoverers, singled out Korea as the `most developed nation in the field of printing.` Woodblock printing was better suited to Chinese characters than movable type, which the Chinese also invented, but which did not replace woodblock printing. In China and Korea, the use of woodblock printing on paper and movable type preceded their use in Europe by several centuries.
Writer	A Writer is anyone who creates a written work, though the word usually designates those who write creatively or professionally, as well as those who have written in many different forms. The word is almost synonymous with author, though somebody who writes, for example, a laundry list, could technically be called the Writer of the list, but not an author. Skilled Writer s are able to use language to portray ideas and images, whether fiction or non-fiction.
Publication	The word Publication means the act of publishing, and it also means any writing of which copies are published, and any website. Among Publications are books, and periodicals, the latter including magazines, scholarly journals, and newspapers. `Publication` is a technical term in legal contexts and especially important in copyright legislation.
Author	An Author (sometimes, in reference to a woman Author, Authoress) is defined both as 'the person who originates or gives existence to anything' and that Authorship determines responsibility for what is created. The second entry goes on to clarify that, when using the term 'Author,' the 'anything' which is created is most usually associated with written work. In copyright law, there is necessarily little flexibility as to what constitutes Authorship.
Editor	The term editor may refer to:

As a person who does editing:

- Editor in chief, having final responsibility for a publication's operations and policies
- Copy editing, making formatting changes and other improvements to text
- Film editing, connecting shots to form a sequence then an entire movie
- Script editor, works with the screenwriter and producer of television dramas and comedies
- Literary editor, in a newspaper or similar, deals with reviews or literary criticism
- Contributing editor, a magazine or book-publishing title, sometimes honorary, with a variety of meanings
- Managing editor, senior member of a publication's management team
- Sound editor, prepares the sound mixing a TV program or film
- Picture editor, collects and reviews photographs and/or illustrations for publication
- Editor at large, a special kind of journalist

As a computer program:

- Character editor, for editing characters in computer games
- Digital audio editor, for editing audio data
- Graphics editor, for creating and manipulating visual images
    - Raster graphics editor, for editing pictures in bitmap or raster format
    - Vector graphics editor, for editing pictures in vector format
- Level editor, for editing levels of computer games
- Hex editor, for editing binary data
- HTML editor, for editing web pages
- MS-DOS Editor, a plain-text editor for MS-DOS and Microsoft Windows
- Source code editor, for editing source code
- Text editor, for editing plain text
- Visual editor, an editing program that displays text onscreen as it is edited
- Word processor, for producing and editing any sort of printable material
- WYSIWYG editor, for editing and visualizing formatted text or graphics
- XML editor, for editing XML data

As other uses:

- Editors, indie rock band from Birmingham, England
- Linear video editing, editing images and sound recorded on videotape
- Non-linear editing system, audio or video editing system allowing random access
- The Editor, villain in the Doctor Who episode "The Long Game"

.

Comma

The comma is a punctuation mark. It has the same shape as an apostrophe or single closing quotation mark in many typefaces, but it differs from them in being placed on the baseline of the text. Some typefaces render it as a small line, slightly curved or straight, or with the appearance of a small, filled-in number 9.

Run-on sentence

A run-on sentence is a sentence in which two or more independent clauses (i.e., complete sentences) are joined without appropriate punctuation or conjunction. It is generally considered to be a grammatical error though it is occasionally used in literature and may be used as a rhetorical device. An example of a run-on is a comma splice, in which two independent clauses are joined with a comma without an accompanying coordinating conjunction.

Pronoun

In linguistics and grammar, a Pronoun is a pro-form that substitutes for a noun with or without a determiner, such as you and they in English. The replaced phrase is called the antecedent of the Pronoun.

For example, consider the sentence `Lisa gave the coat to Phil.` All three nouns in the sentence can be replaced by Pronouns: `She gave it to him.` If the coat, Lisa, and Phil have been previously mentioned, the listener can deduce what the Pronouns she, it and him refer to and therefore understand the meaning of the sentence; however, if the sentence `She gave it to him.` is the first presentation of the idea, none of the Pronouns have antecedents, and each Pronoun is therefore ambiguous.

Conjunction

In grammar, a conjunction is a part of speech that connects two words, phrases or clauses together. This definition may overlap with that of other parts of speech, so what constitutes a 'conjunction' should be defined for each language. In general, a conjunction is an invariable grammatical particle, and it may or may not stand between the items it conjoins.

Dependent clause

A Dependent clause cannot stand alone as a sentence. A Dependent clause is usually attached to an inDependent clause. Although a Dependent clause contains a subject and a predicate, it sounds incomplete when standing alone.

Modifier	In grammar, a modifier is an optional element in phrase structure or clause structure; the removal of the modifier typically doesn`t affect the grammaticality of the construction. modifiers can be a word, a phrase or an entire clause. Semantically, modifiers describe and provide more accurate definitional meaning for another element.
Parallelism	In grammar, parallelism is a balance of two or more similar words, phrases, or clauses. The application of parallelism in sentence construction improves writing style and readability. parallelism may also be known as parallel structure or parallel construction.
Punctuation	Punctuation marks are symbols that indicate the structure and organization of written language, as well as intonation and pauses to be observed when reading aloud. In written English, punctuation is vital to disambiguate the meaning of sentences. For example, "woman, without her man, is nothing" and "woman: without her, man is nothing" have greatly different meanings, as do "eats shoots and leaves" and "eats, shoots and leaves".
Question mark	The question mark , is a punctuation mark that replaces the full stop (period) at the end of an interrogative sentence in English and many other languages. The question mark is not used for indirect questions. The question mark character is also often used in place of missing or unknown data.
Apostrophe	The apostrophe is a punctuation mark, and sometimes a diacritic mark, in languages that use the Latin alphabet or certain other alphabets. In English, it serves three purposes: • The omission of one or more letters (as in the contraction of do not to don't). • The marking of possessive case (as in the cat's whiskers). • The marking as plural of written items that are not words established in English orthography (as in P's and Q's, the late 1950's). (This is considered incorrect by some.

Colon	The Colon is a punctuation mark consisting of two equally sized dots centered on the same vertical line. Punctuation Usage As with many other punctuation marks, the usage of colon varies among languages and, within a given language, across historical periods. As a rule, however, a colon informs the reader that the following proves, explains or simply provides elements of what is referred to before.
Semicolon	The Semicolon is a punctuation mark with several uses. The Italian printer Aldus Manutius the Elder established the practice of using the Semicolon mark to separate words of opposed meaning, and to indicate interdependent statements. The earliest, general use of the Semicolon in English was in 1591; Ben Jonson was the first notable English writer to use them systematically.
Comma	The comma is a punctuation mark. It has the same shape as an apostrophe or single closing quotation mark in many typefaces, but it differs from them in being placed on the baseline of the text. Some typefaces render it as a small line, slightly curved or straight, or with the appearance of a small, filled-in number 9.
Apostrophe	The apostrophe is a punctuation mark, and sometimes a diacritic mark, in languages that use the Latin alphabet or certain other alphabets. In English, it serves three purposes: • The omission of one or more letters (as in the contraction of do not to don't). • The marking of possessive case (as in the cat's whiskers). • The marking as plural of written items that are not words established in English orthography (as in P's and Q's, the late 1950's). (This is considered incorrect by some.

Bracket

Brackets are tall punctuation marks used in matched pairs within text, to set apart or interject other text. In the United States, "bracket" usually refers specifically to the "square" or "box" type; in British usage it normally refers to the "round" type, which is called a parenthesis mark in American usage.

List of types

- round brackets, open brackets, brackets (UK), or parentheses: ( )
- square brackets, closed brackets, or brackets (US): [ ]
- curly brackets, definite brackets, swirly brackets, birdie brackets, Scottish brackets, squirrelly brackets, braces, gullwings, chicken lips, fancy brackets, or squiggly brackets: { }
- angle brackets, triangular brackets, diamond brackets, tuples, or chevrons: ? ?
- inequality signs: < >
- corner brackets: ? ?; ? ?

History

The chevron was the earliest type to appear in written English.

Transition

Transitions in fiction are words, phrases, sentences, paragraphs, or punctuation that may be used to signal various changes in a story, including changes in time, location, point-of-view character, mood, tone, emotion, and pace. Transitions are sometimes listed as one of various fiction-writing modes.

Purpose of transitions

Transitions provide for a seamless narrative flow as a story shifts in time, location, or point-of view.

Synonyms

Synonyms are different words with identical or very similar meanings. Words that are Synonyms are said to be synonymous, and the state of being a synonym is called synonymy. The word comes from Ancient Greek syn (`with`) and onoma (á½„νομα) (`name`).

Phrase	In grammar, a Phrase is a group of words functioning as a single unit in the syntax of a sentence. For example, the store at the end of the street is a Phrase. It acts like a noun.
Writing process	The Writing process is both a key concept in the teaching of writing and an important research concept in the field of composition studies. Research on the writing process focuses on how writers draft, revise, and edit texts. Composing process research was pioneered by scholars such as Janet Emig in "The Composing Processes of Twelfth Graders" (1971), Sondra Perl in "The Composing Processes of Unskilled College Writers (1979), and Linda Flower and John R. Hayes in "A Cognitive Process Theory of Writing" (1981).
Brainstorming	Brainstorming is a group creativity technique designed to generate a large number of ideas for the solution of a problem. In 1953 the method was popularized by Alex Faickney Osborn in a book called Applied Imagination. Osborn proposed that groups could double their creative output with Brainstorming.
Revision	Revision is the stage in the writing process where the author reviews, alters, and amends her or his message, according to what has been written in the draft. Revision follows drafting and precedes editing. Drafting and revising often form a loop as a work moves back and forth between the two stages.
Photograph	A Photograph is an image created by light falling on a light-sensitive surface, usually Photographic film or an electronic imager such as a CCD or a CMOS chip. Most Photographs are created using a camera, which uses a lens to focus the scene`s visible wavelengths of light into a reproduction of what the human eye would see. The process and practice of creating Photographs is called Photography.

CPSIA information can be obtained at www.ICGtesting.com
231678LV00001B/13/P

9 781614 905554